STATISTICS
WITHOUT FORMULAS

A Cartoon Book

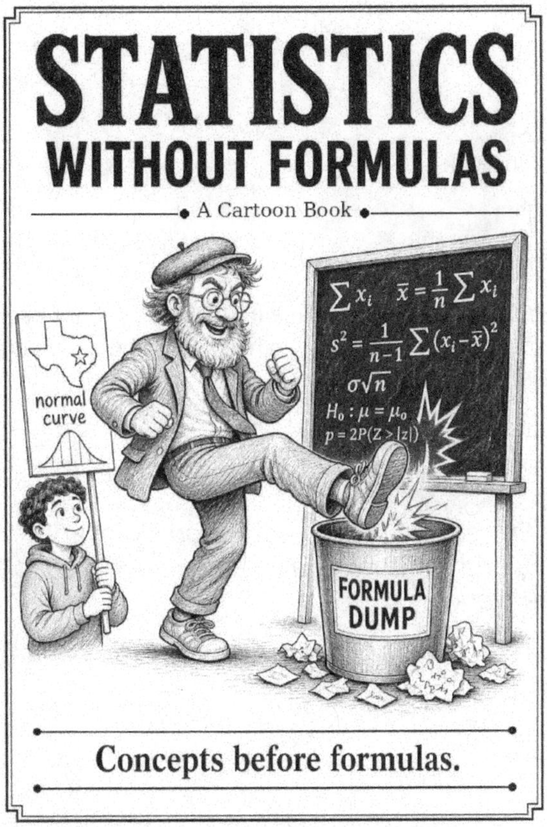

Michael M. Nikoletseas, Ph.D.

Concepts before formulas

Statistics Without Formulas: A Cartoon Book

Michael M. Nikoletseas, Ph.D.

Published in the United States of America.

ISBN: 9798182736312

All cartoons were created for this edition. The order of topics follows the concept-first teaching of Michael M. Nikoletseas, Ph.D., and his book Statistics for High School Students (revised August 2026). This book is a separate work.

Each topic begins with a picture instead of a formula. Study the picture, describe what is happening, and then read the explanation.

Foreword

Statistics is often taught by starting with formulas. This book starts with the ideas. It shows what each statistical procedure does before asking the reader to calculate. Most parametric methods measure variability and compare one source of variability with another.

I am a retired professor of Medicine and Neuroscience. I taught laboratory courses and statistics for many years. A beginning student can understand the main ideas with addition, subtraction, multiplication, and division. The book explains levels of measurement, measures of center, variability, observed distributions, the ideal normal curve, the t distribution, and basic experimental designs.

The book moves from the mean to mixed and split-plot designs. The symbols t, F, SS, and MS are introduced only after the actions they represent are clear.

Data may arrive as words, numbers, or images. First identify the type of data and organize it. Then find a center, draw a graph, and measure the spread. Biological measurements may form an observed distribution. The normal curve is a separate ideal model measured in standard-deviation units. When variability is estimated from a small sample, the t distribution has heavier tails. As the degrees of freedom increase, the t distribution approaches the normal distribution. In the opening chalkboard, σ denotes the population standard deviation; s denotes a standard deviation calculated from a sample.

Begin with the cartoon. Describe the action in plain words before reading the explanation.

Michael M. Nikoletseas

Table of Contents

How to use this book

The chapters follow a simple order. Begin with unorganized data, identify the level of measurement, organize the data, find a center, draw a graph, and measure variability. Next, compare an observed distribution with the ideal normal curve. The t distribution is used when population variability must be estimated from a sample. At about 50 degrees of freedom, it is very close to the normal distribution. The final chapters introduce one-way, factorial, repeated-measures, and mixed designs.

Begin with the cartoon. Describe the action in plain words before reading the explanation.

PART I
From data to description

Unorganized data

Figure 1. Unorganized data

Scientific data often arrive in several forms: notes, photographs, labels, measurements, and counts. Before performing a statistical test, identify what each item represents and organize the data into a consistent form.

Data and levels of measurement

Figure 2. Four scales

Not every number measures an amount. A number may be a name, a rank, an interval value, or a ratio value. Use the highest level of measurement that the data truly support. Means, standard deviations, t tests, and F tests are usually used with interval or ratio data. Nominal and ordinal data usually require other summaries and tests.

Nominal

Numbers used as names

Figure 3. Counting

A count tells how many items are present. It does not measure height, distance, or quality.

Jersey numbers

Figure 4. Jersey numbers

Jersey number 11 is not a larger amount than jersey number 3. These numbers identify players; they do not measure quantity.

Ordinal

Rank, not amount

Figure 5. Ordinal ranks in a race

First, second, and third show order. They do not show the time difference between runners.

Interval and ratio scales

Figure 6. Interval versus ratio

An interval scale has equal units. A ratio scale also has a meaningful absolute zero. For example, zero kelvin is absolute thermodynamic zero, so kelvin temperatures may be compared as ratios. Statements such as 'twice as much' require a ratio scale.

First step: organize the data

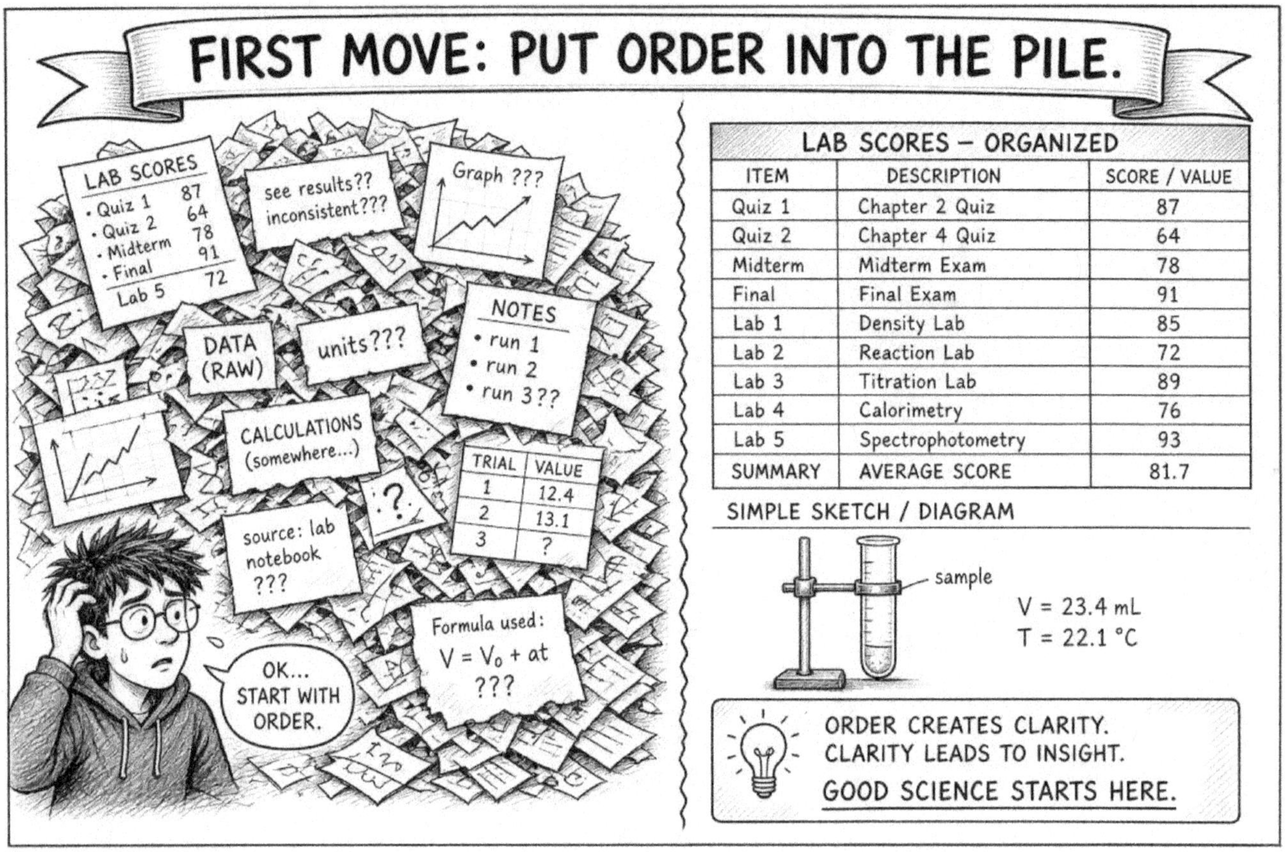

Figure 7. Organizing data

A table identifies which values belong together and what each row and column represents. This organization must be correct before calculating a mean, drawing a graph, or performing a statistical test.

The mean

Figure 8. Calculating the mean

The mean is one measure of the center of a set of scores. Add all scores and divide by the number of scores. For 2, 4, 6, 8, and 10, the total is 30 and the mean is 6. Because every score contributes to the mean, an extreme score can change it greatly.

This simple arithmetic operation can also be written in formula form.

$$\text{mean} = \frac{\text{sum of all sample scores}}{\text{number of scores}}$$

$$\bar{x} = \frac{\sum x}{n}$$

$$\bar{x} = \frac{1}{n}\sum_{i=1}^{n} x_i$$

Median and mode

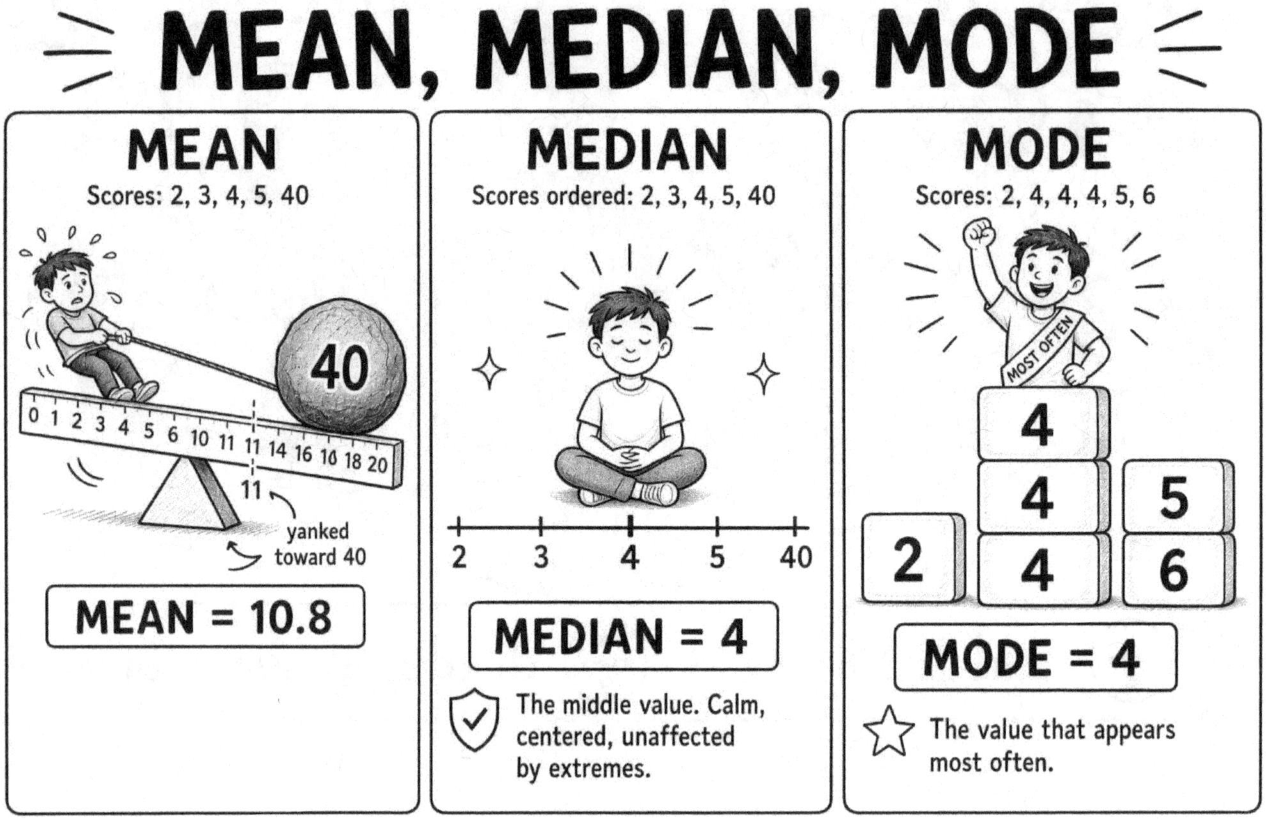

Figure 9. Mean, median, and mode

Arrange the scores from smallest to largest. In 2, 3, 4, 5, and 40, the mean is 10.8 because the extreme score 40 raises it. The median is 4, the middle value. The mode is the most frequent value; in 2, 4, 4, 4, 5, and 6, the mode is 4.

Graphs

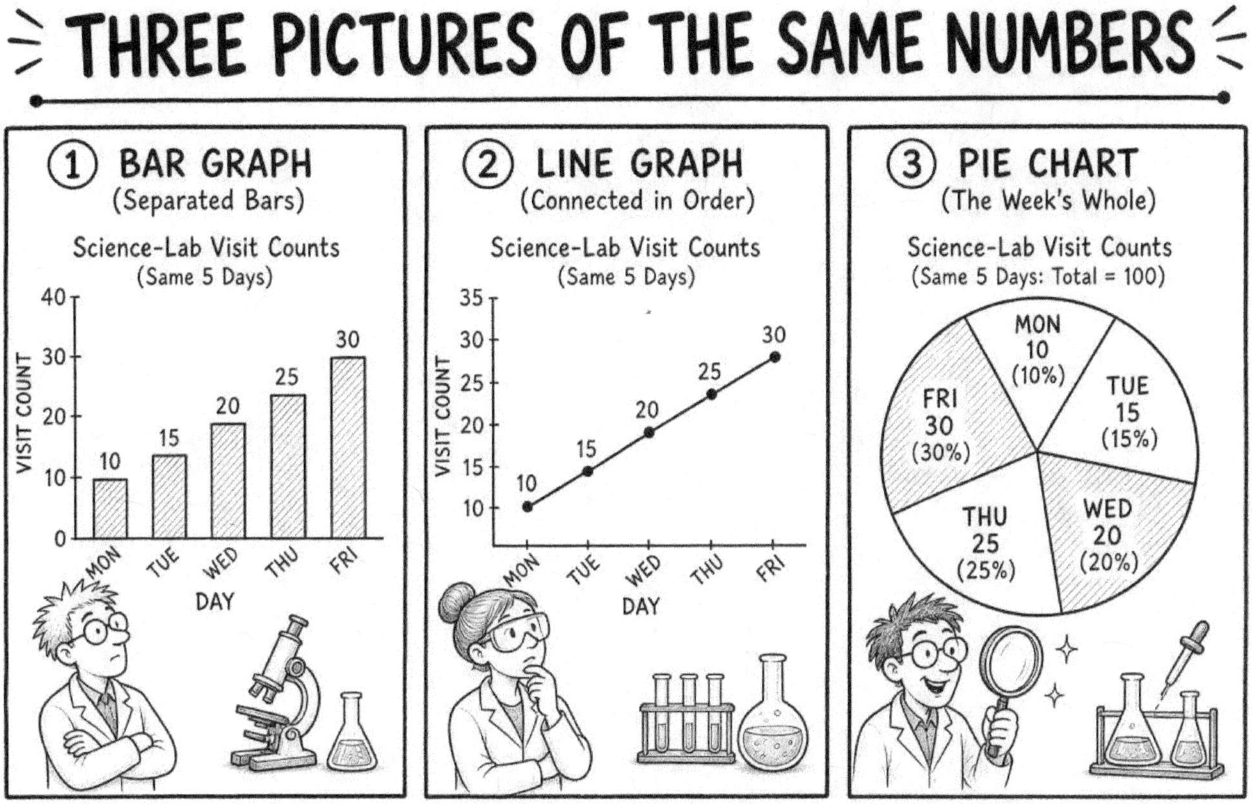

Figure 10. Bar, line, and pie graphs

A graph displays numerical patterns. Bar graphs compare separate categories. Line graphs show ordered observations, such as changes across days. Pie graphs show parts of one whole. Scatterplots show two measurements for each subject.

PART II

From spread to variance

Understanding variability

Same mean, different spread

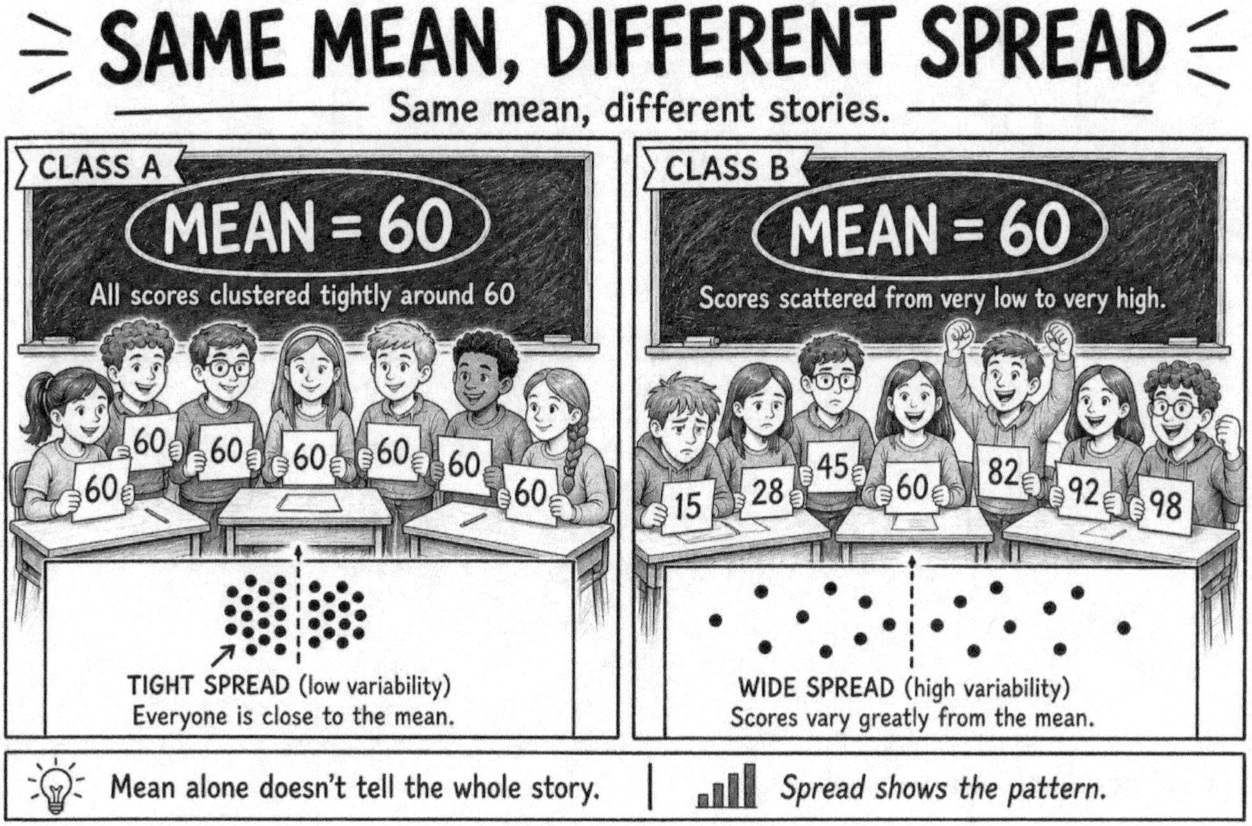

Figure 11. Same mean different spread

Two classes can have the same mean of 60 but very different score distributions. One class may have scores close to 60, while another ranges from 15 to 98. The mean does not describe this difference. Variability measures how far scores are from the center.

Distances from the mean

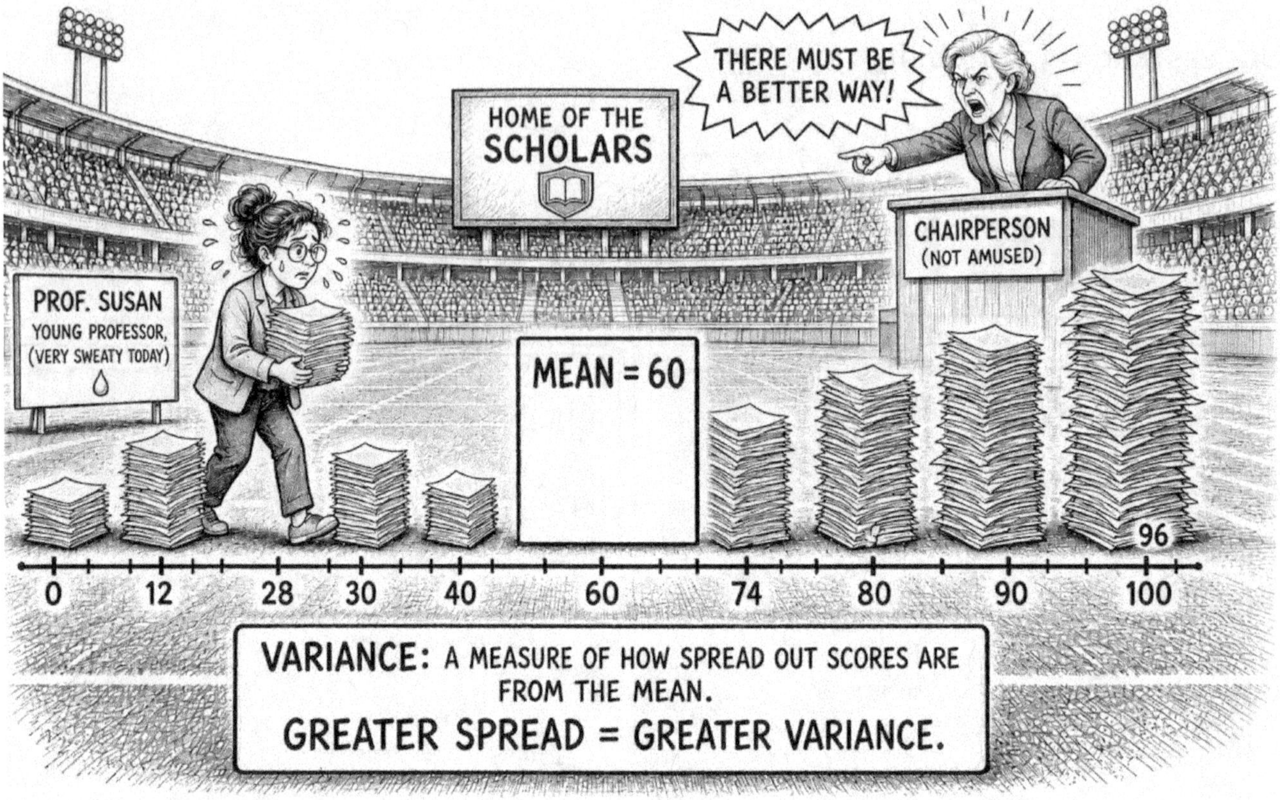

Figure 12. Showing distances from the mean

The distance from each score to the mean shows how spread out the scores are. Variance records and combines these distances numerically.

This simple, intuitive idea can also be written in formula form.

$$\text{sample variance} = \frac{\text{sum of squared distances from the sample mean}}{\text{number of sample scores} - 1}$$

$$s^2 = \frac{\sum (x - \bar{x})^2}{n - 1}$$

$$s^2 = \frac{1}{n-1} \sum_{i=1}^{n} (x_i - \bar{x})^2$$

Variance and the sum of squares - From SS to MS

Figure 13. From sum of squares to mean square

Variance summarizes squared distances from the mean. Standard deviation returns the measure to the original units.

Find the distance between each score and the mean. Positive and negative distances would add to zero, so square each distance. Add the squared distances. The result is the sum of squares, or SS.

For a sample, divide SS by the degrees of freedom. One degree of freedom is lost when the sample mean is calculated, so the divisor is n minus 1. The result is a mean square, or MS, and it is the sample variance. The standard deviation is the square root of the variance.

SS totals squared distances. MS divides SS by degrees of freedom. SD returns to the original scale.

$$\text{sum of squares} = \text{sum of squared distances from the mean}$$

$$\text{mean square} = \frac{\text{sum of squares}}{\text{degrees of freedom}}, \qquad \text{standard deviation} = \sqrt{\text{mean square}}$$

$$SS = \sum (x - \bar{x})^2, \qquad MS = \frac{SS}{df}, \qquad s = \sqrt{MS}$$

$$SS = \sum_{i=1}^{n} (x_i - \bar{x})^2, \qquad MS = \frac{SS}{n-1}, \qquad s = \sqrt{\frac{1}{n-1}\sum_{i=1}^{n} (x_i - \bar{x})^2}$$

PART III

From the normal curve to inference

Height measurements form a data curve

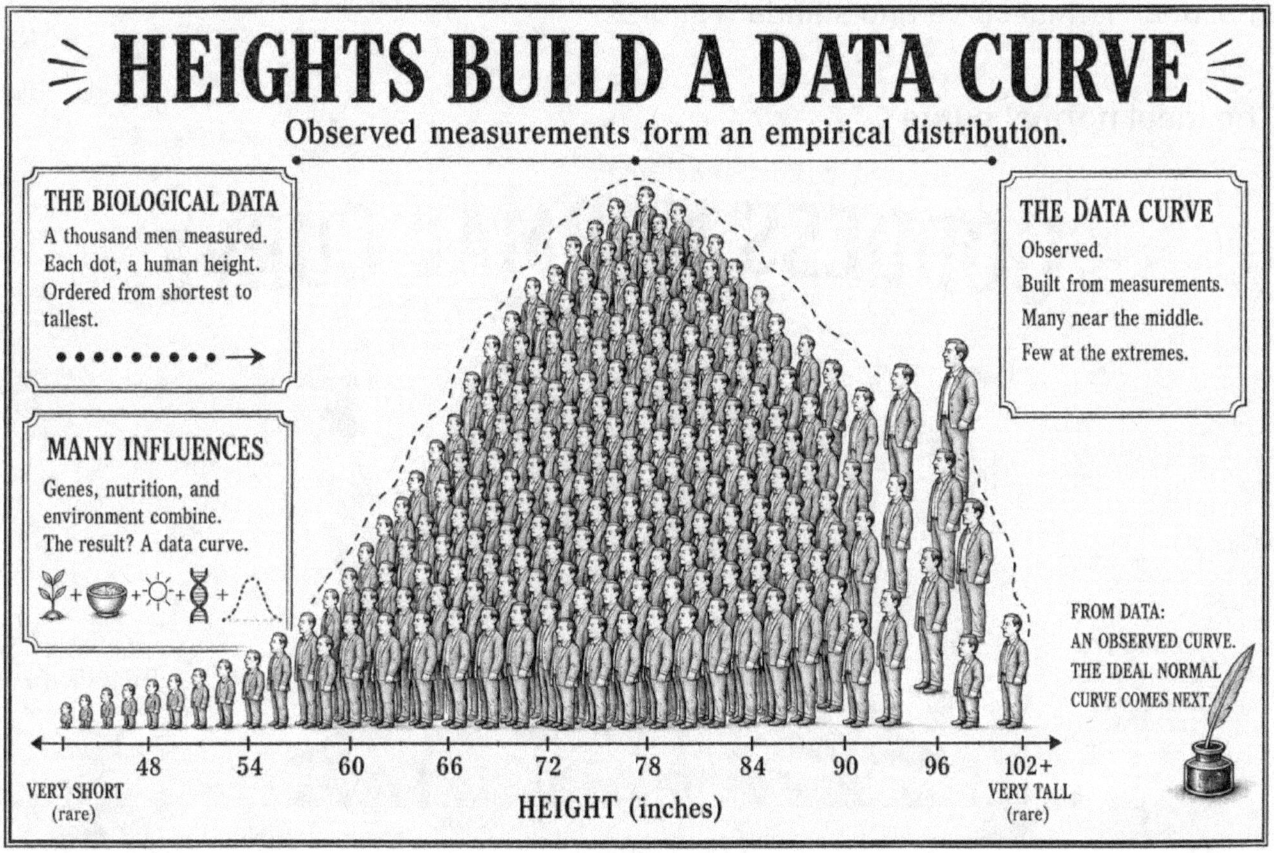

Figure 14. A data curve formed by height measurements

This curve comes from observed measurements. The normal curve introduced next is an ideal model.

Measure the heights of a thousand men and arrange them from shortest to tallest. Few measurements occur at the extremes, and many occur near the center. These observations form a data distribution. It may resemble a normal curve, but the normal curve is a separate theoretical model.

The ideal normal curve and standard scores

The ideal normal curve

Figure 15. The ideal normal curve

The normal curve is a theoretical model. Real data may approximate it, but no observed distribution is perfectly normal. Its fixed proportions make it useful as a reference.

Constructing the Z table

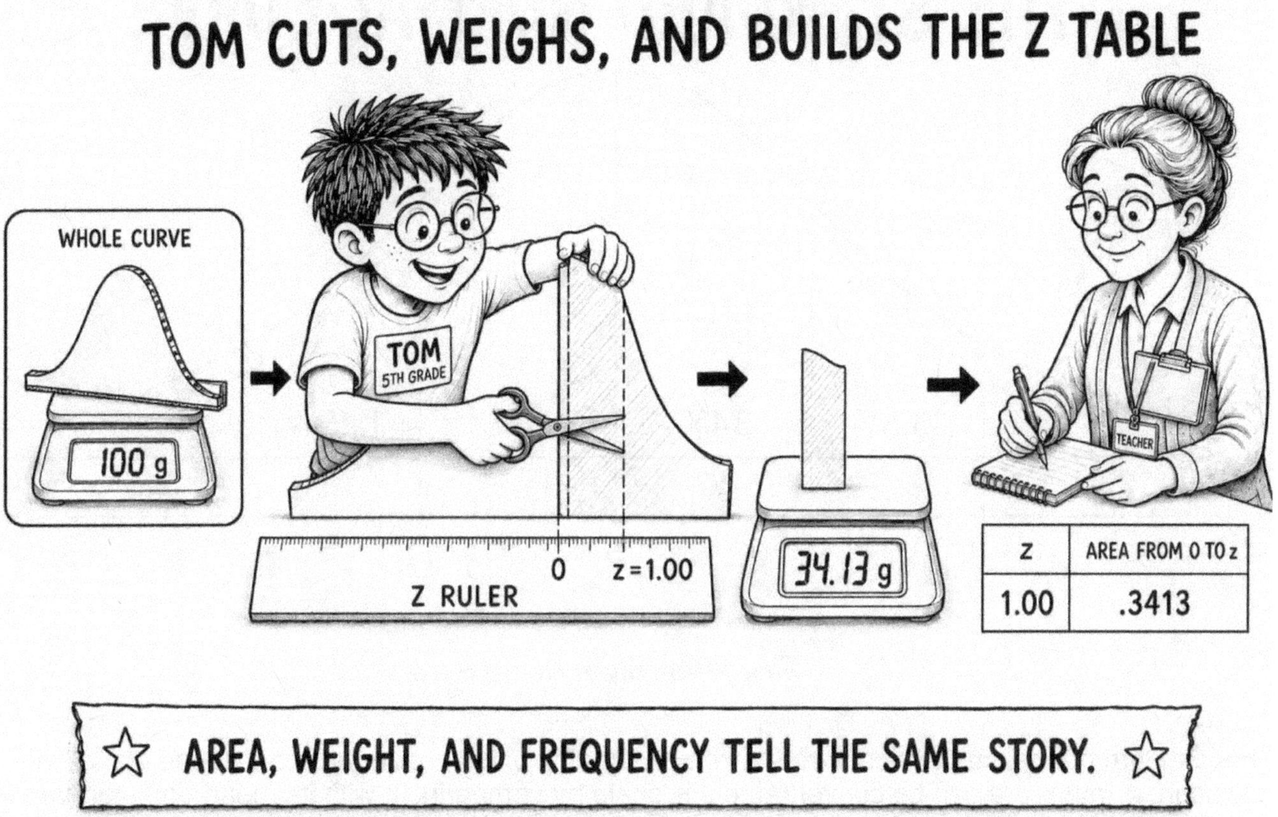

Figure 16. Constructing the Z table by cutting and weighing

The Z table connects a position on the normal curve with the proportion of area at that position.

Tom cuts the ideal normal curve from cardboard of uniform thickness. The whole curve weighs 100 grams. He cuts the section from z = 0 to z = 1.00 and weighs it. The section weighs 34.13 grams, so it represents 34.13 percent of the total area. On the normal curve, this area is the expected proportion of observations in that interval. The Z table records z = 1.00 and area = .3413.

Standard normal curve - standard scores

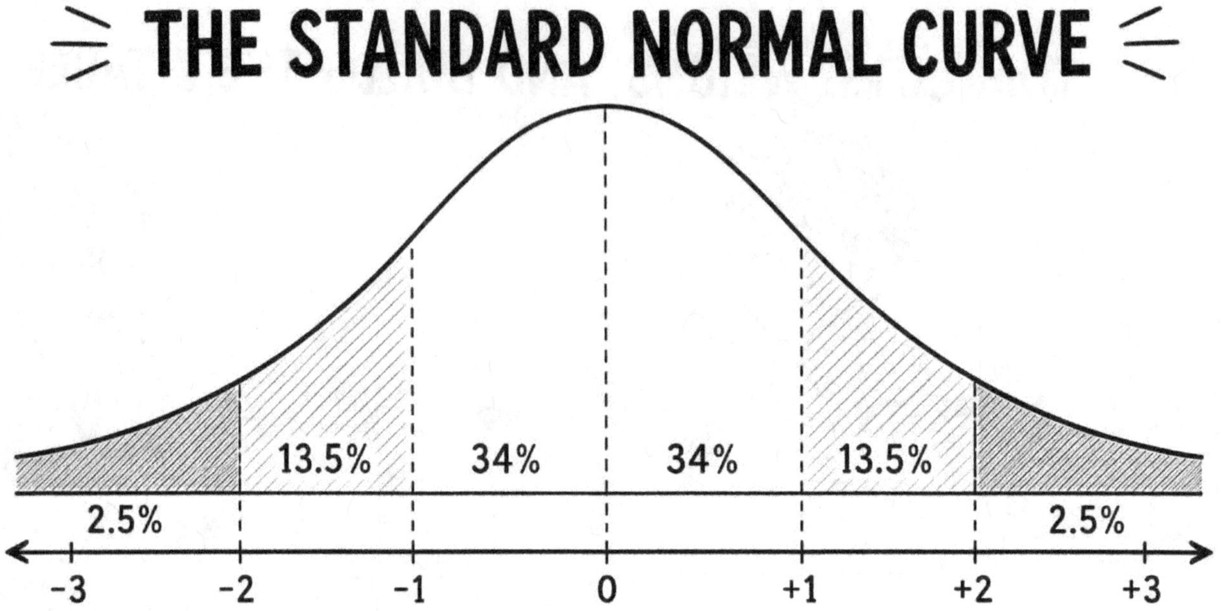

Figure 17. Standard normal curve

The standard normal curve is centered at zero. Each unit on its horizontal scale is one standard deviation. A raw score can be converted to this scale by comparing it with its mean and standard deviation. The resulting position shows how many standard deviations the score is above or below the mean.

$$\text{standard score} = \frac{\text{score} - \text{sample mean}}{\text{sample standard deviation}}$$

$$z = \frac{x - \bar{x}}{s}$$

$$z_i = \frac{x_i - \bar{x}}{s}$$

Four uses of the normal curve

Locate a score

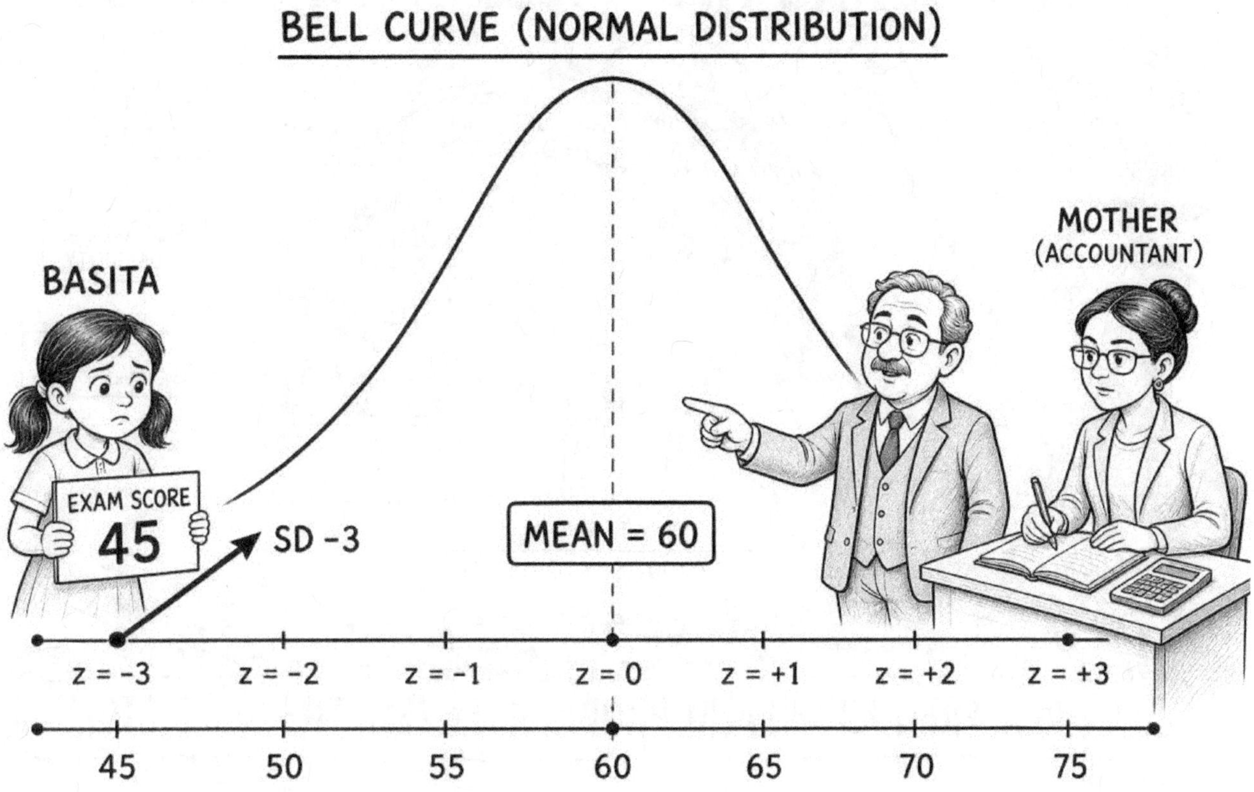

Figure 18. A score below the mean

If the mean is 60 and the standard deviation is 5, a raw score of 45 lies exactly three standard deviations below the mean. Most scores in a normal distribution are above that position.

Estimate the probability of a new observation

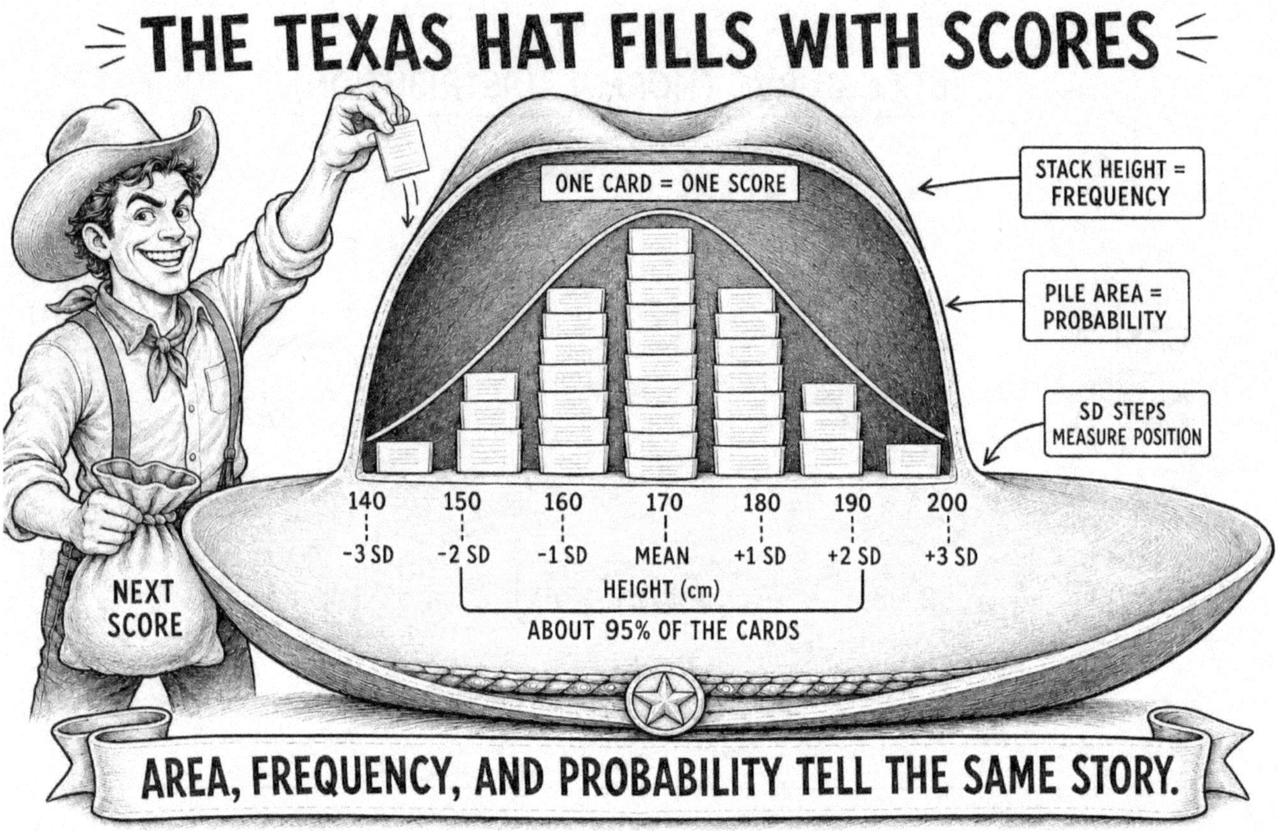

Figure 19. Scores fill the Texas normal-curve hat

Each height measurement is placed at its position on the scale. Equal values stack together. Few measurements occur at the extremes, and most occur near 170 cm. The height of each stack shows frequency. The total area in a region represents probability. The interval from 150 to 190 cm is within two standard deviations of the mean and contains about 95 percent of a normal distribution.

Estimate a population mean

Figure 20. Sample size and standard error

Standard error describes the spread and variability of sample means.

Larger samples give smaller standard errors and more precise estimates, but they do not guarantee an unbiased or representative sample.

$$\text{standard error of the mean} = \frac{\text{sample standard deviation}}{\sqrt{\text{sample size}}}$$

$$\text{SE} = \frac{s}{\sqrt{n}}$$

$$\text{SE}_{\bar{x}} = \frac{s}{\sqrt{n}}$$

Compare two means

The Two-Sample Z-Test

The two-sample Z-test compares the means of two independent samples when both population standard deviations are known. It divides the difference between the sample means by the standard error of that difference. The resulting Z-score shows how far the observed difference lies from the hypothesized difference of zero.

Simple form

$$\text{Z-score} = \frac{\text{difference between the two sample means}}{\text{standard error of the difference}}$$

Symbolic form

$$Z = \frac{(\bar{X}_1 - \bar{X}_2) - 0}{\sqrt{\frac{\sigma_1^2}{n_1} + \frac{\sigma_2^2}{n_2}}}$$

Here, $\bar{X}_1$ and $\bar{X}_2$ are the sample means, σ_1 and σ_2 are the known population standard deviations, and n_1 and n_2 are the sample sizes.

Complex expanded form

$$Z = \frac{\left(\frac{1}{n_1}\sum_{i=1}^{n_1} x_{1i} - \frac{1}{n_2}\sum_{j=1}^{n_2} x_{2j}\right) - 0}{\sqrt{\frac{\sigma_1^2}{n_1} + \frac{\sigma_2^2}{n_2}}}$$

Here, x_{1i} represents each score in the first sample, and x_{2j} represents each score in the second sample.

$$t = \frac{\text{difference between the sample means}}{\text{standard error of the difference}}$$

$$t = \frac{\bar{x}_1 - \bar{x}_2}{\mathrm{SE}_{\bar{x}_1 - \bar{x}_2}}$$

$$t = \frac{\bar{x}_1 - \bar{x}_2}{s_p\sqrt{\frac{1}{n_1} + \frac{1}{n_2}}}, \qquad s_p^2 = \frac{(n_1 - 1)s_1^2 + (n_2 - 1)s_2^2}{n_1 + n_2 - 2}$$

From the z-table to the t-table

Figure 21. From the Z table to the t table

The Z table gives areas for one fixed normal curve. The t table uses a family of curves. First determine the degrees of freedom from the sample information. For two independent samples, account for the mean calculated in each group. Then select the appropriate row of the t table and the required tail probability.

The t distribution

How the t distribution differs from the normal distribution

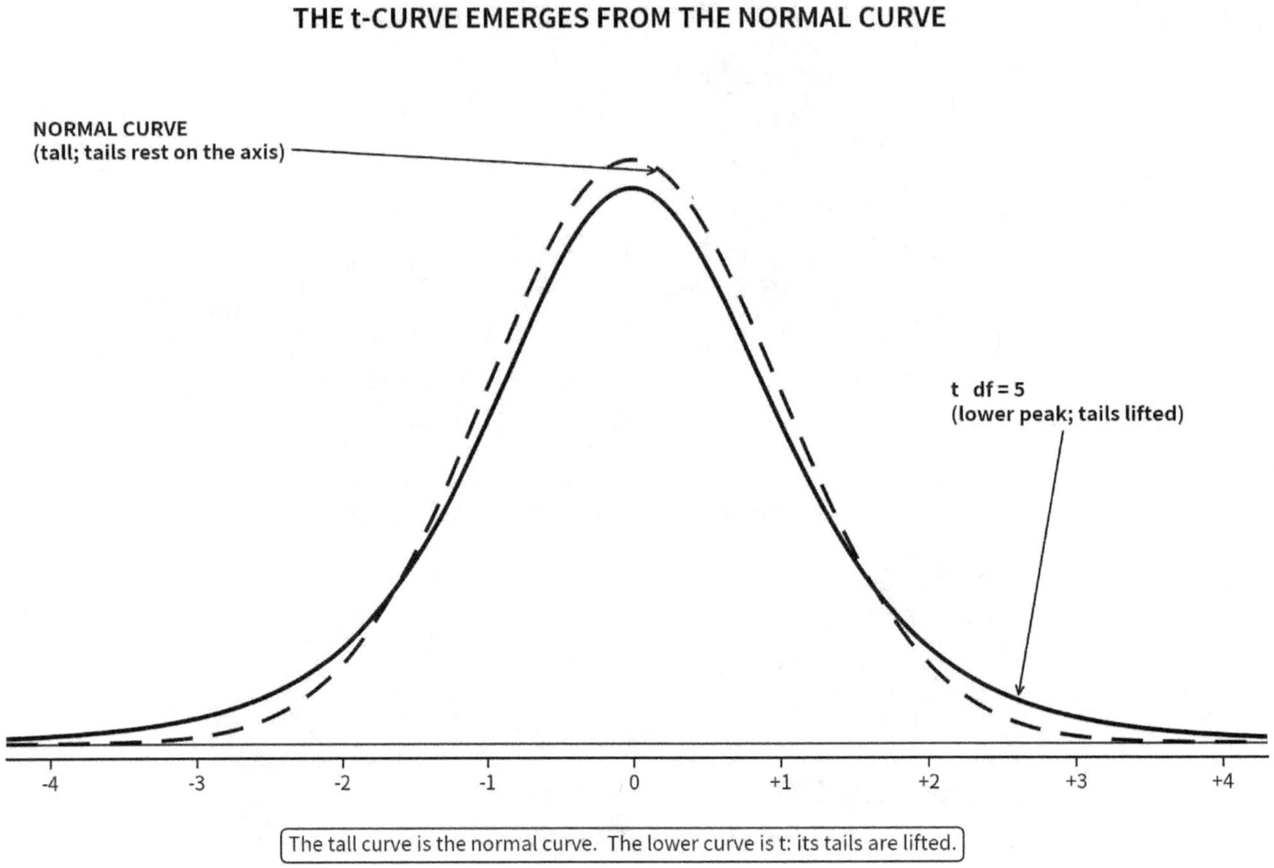

Figure 22. Normal and t distributions

The t distribution is centered at zero like the normal distribution, but it has a lower peak and heavier tails. Its shape depends on the degrees of freedom. To use a t table, select the degrees-of-freedom row and the required tail-probability column. Their intersection gives the critical t value. As the degrees of freedom increase, the t distribution approaches the normal distribution.

Heavier tails contain more area

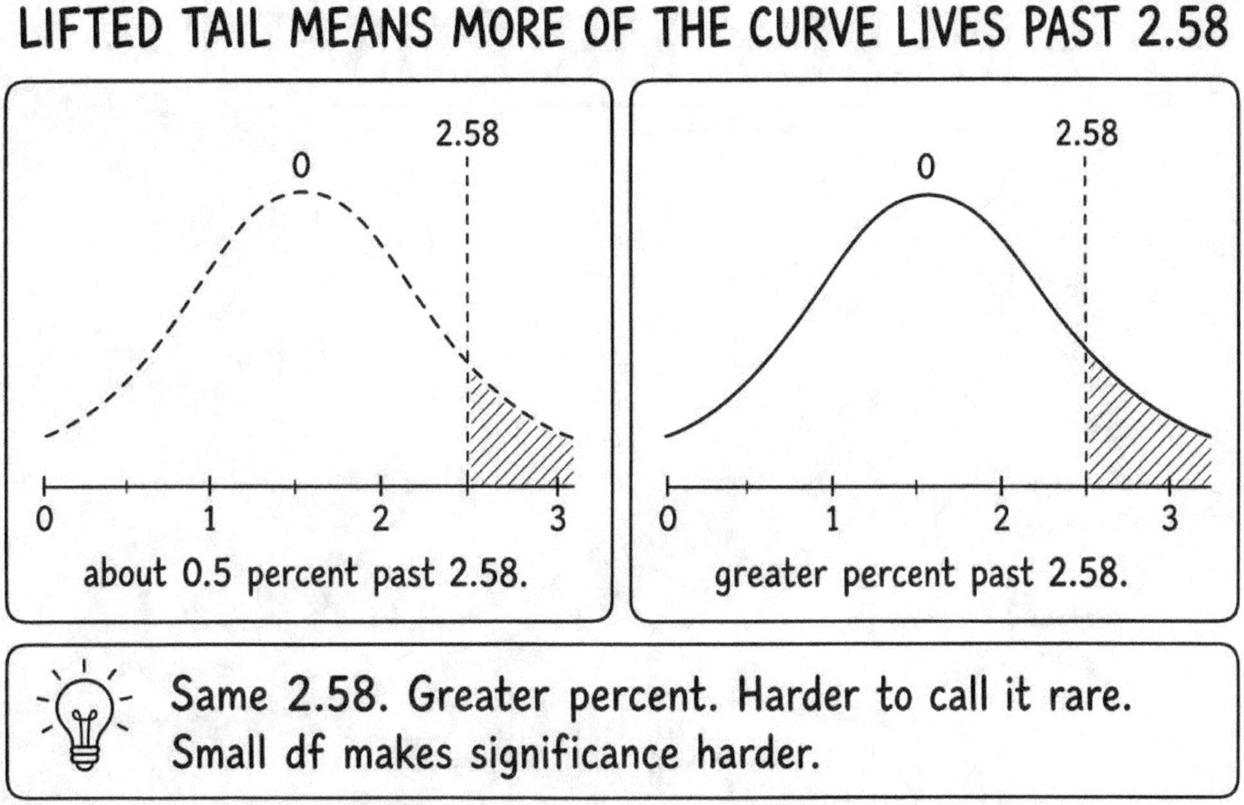

Figure 23. Heavier tails place more area beyond 2.58

A t distribution with few degrees of freedom places less area near the center and more area in the tails. Therefore, a value of 2.58 is less unusual under that t distribution than under the normal distribution. A larger absolute t value is required to reach a tail probability of 5 percent or 1 percent. Smaller samples require larger critical values to achieve statistical significance.

Small samples and the t table

Figure 24. The t table for small samples

With a small sample, an observed difference must be larger to be considered unusual. The t table gives the required critical value for each number of degrees of freedom.

Why few degrees of freedom require a larger critical value

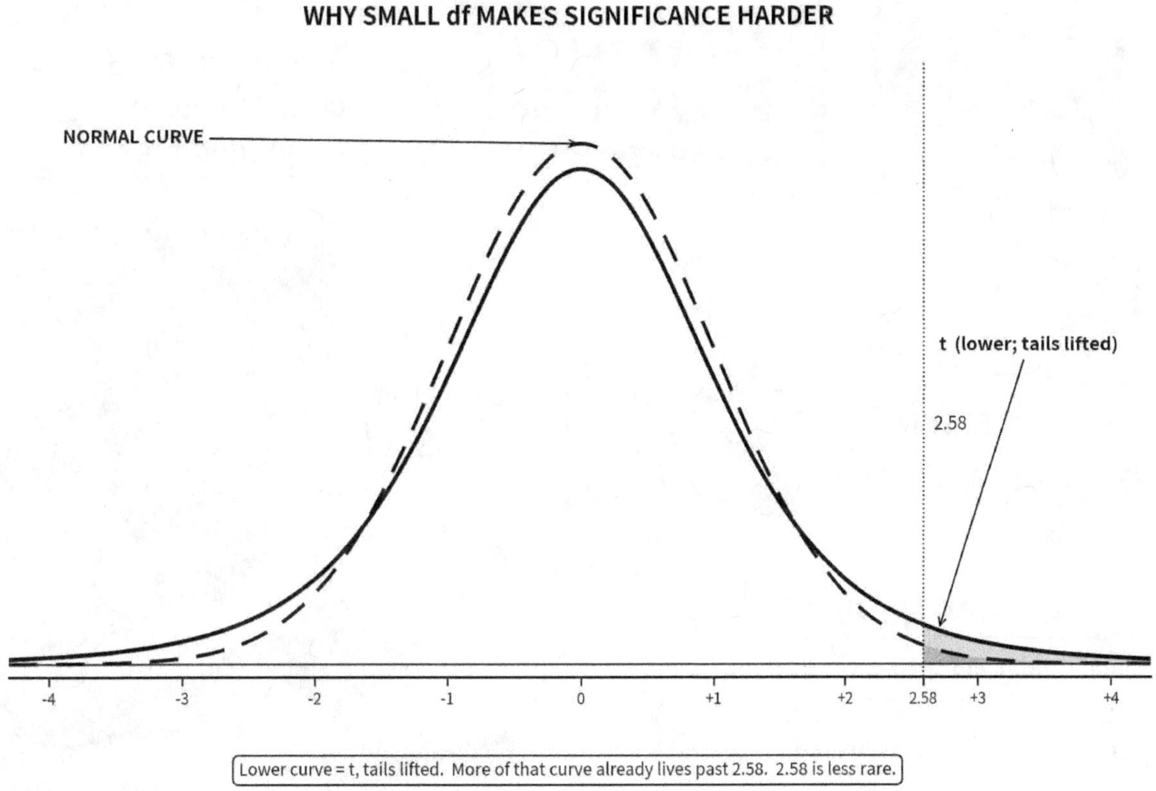

Figure 25. Critical values with few degrees of freedom

Both curves are centered at zero. Beyond 2.58, a t distribution with few degrees of freedom has more area than the normal distribution. The same observed value therefore has a larger p value under the t distribution. To obtain the same tail probability, the critical t value must be farther from zero.

Degrees of freedom

Figure 26. Degrees of freedom after calculating a mean

Degrees of freedom count the independent pieces of information left after estimating the means. With two samples of six, 5 + 5 = 10.

$$\text{degrees of freedom} = (\text{first sample size} - 1) + (\text{second sample size} - 1)$$

$$df = (n_1 - 1) + (n_2 - 1)$$

$$df = (n_1 - 1) + (n_2 - 1) = n_1 + n_2 - 2$$

The next designs compare variability associated with an effect against error variability.

PART IV

Comparing sources of variability

Statistical tests

Analysis of Variance (ANOVA)

Comparing several means

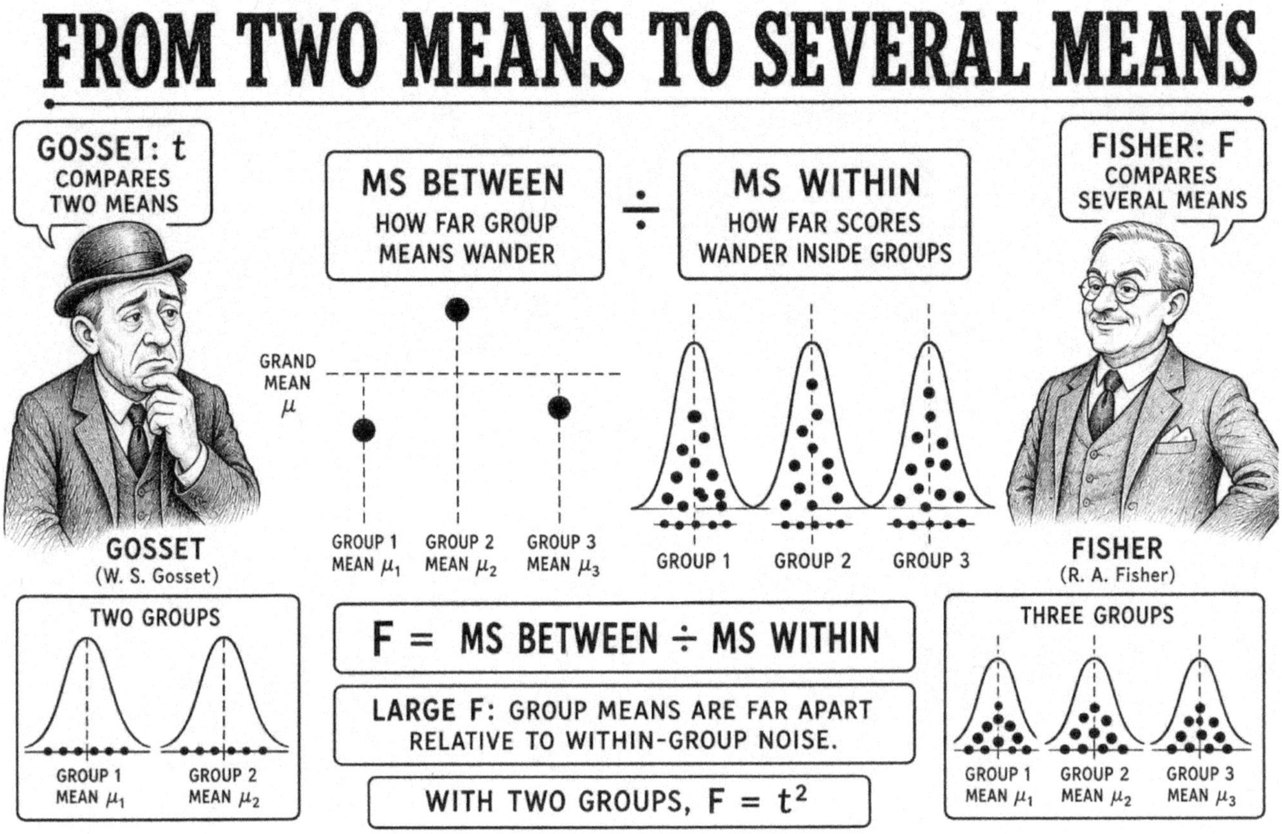

Figure 27. Comparing several means

One-way design

Figure 28. One-way layout

A one-way design has one factor and two or more independent groups. Each subject receives only one level of the factor. The analysis compares variability among the group means with variability among subjects within the groups.

Between-group and within-group variability

Figure 29. Between-group and within-group variability

Between-group variability describes how far the group means are from the grand mean. Within-group variability describes differences among subjects in the same group. The F ratio compares these two sources of variability. A significant F indicates that at least one group mean differs, but it does not identify which groups differ.

$$F = \frac{\text{variability between groups}}{\text{variability within groups}}$$

$$F = \frac{MS_B}{MS_W}$$

$$F = \frac{\dfrac{\sum_{j=1}^{k} n_j(\bar{x}_j - \bar{x}_{..})^2}{k-1}}{\dfrac{\sum_{j=1}^{k}\sum_{i=1}^{n_j}(x_{ij} - \bar{x}_j)^2}{N-k}}$$

Factorial designs

Two-by-two factorial design

Figure 30. Two-by-two factorial design

A two-by-two factorial design has two factors, each with two levels. All four combinations are included, and each subject belongs to one combination. A main effect tests one factor while averaging across the other factor. An interaction tests whether the effect of one factor changes across the levels of the other. Nonparallel lines show this change.

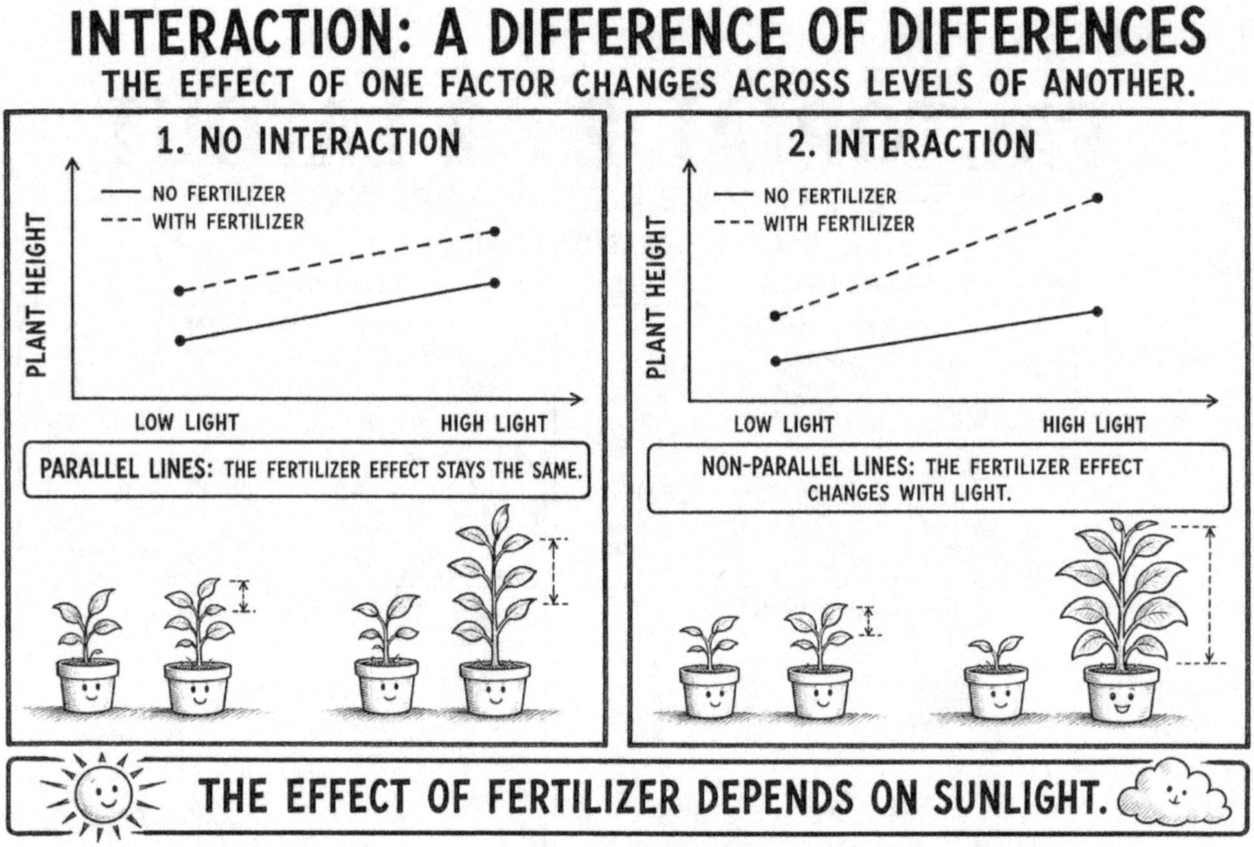

Figure 31. Interaction

Compare the fertilizer effect under low and high light. Under low light, fertilizer adds little height. Under high light, it adds much more. The effect of fertilizer therefore depends on the light level. Equal gaps produce parallel lines and no interaction. Unequal gaps produce nonparallel lines and an interaction.

Repeated measures

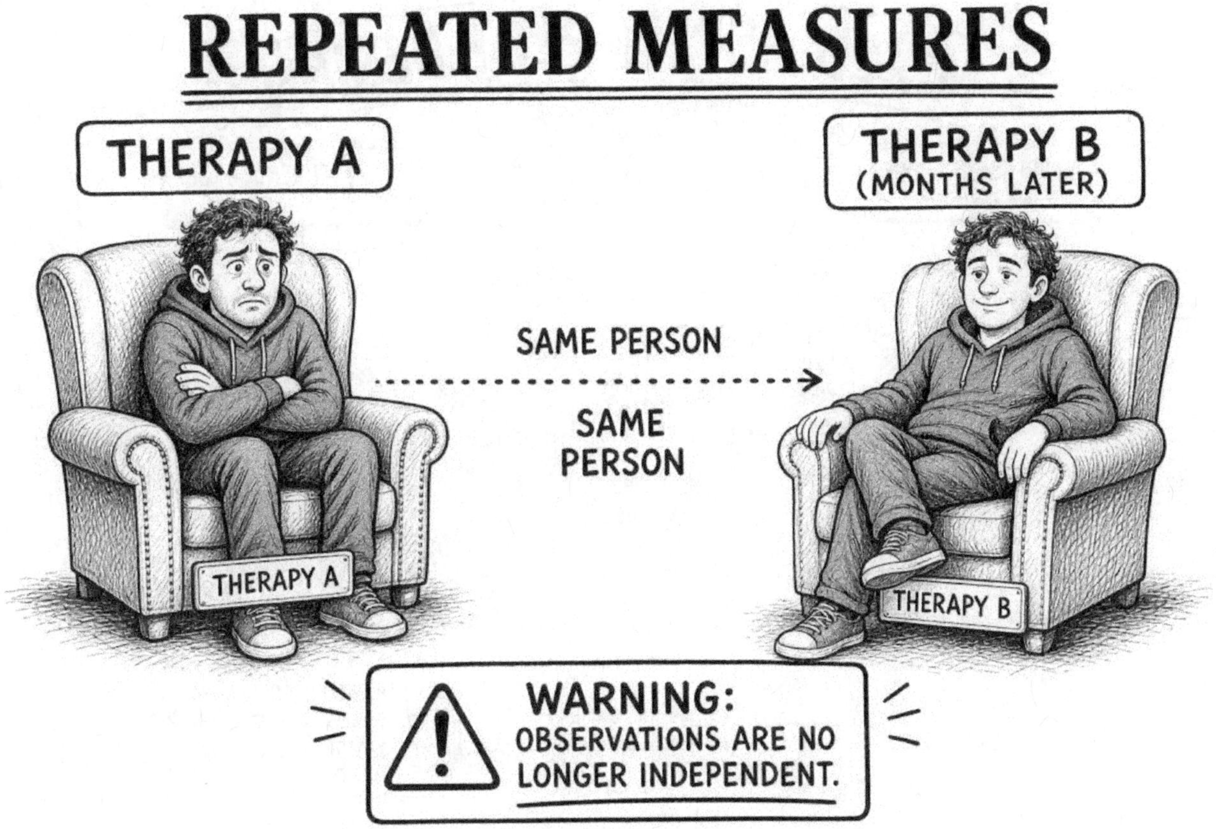

Figure 32. Repeated measures

In a repeated-measures design, the same subject is measured more than once. The observations are related because they come from the same person. This design controls for differences among subjects, but an earlier treatment may influence a later measurement.

Mixed split-plot design

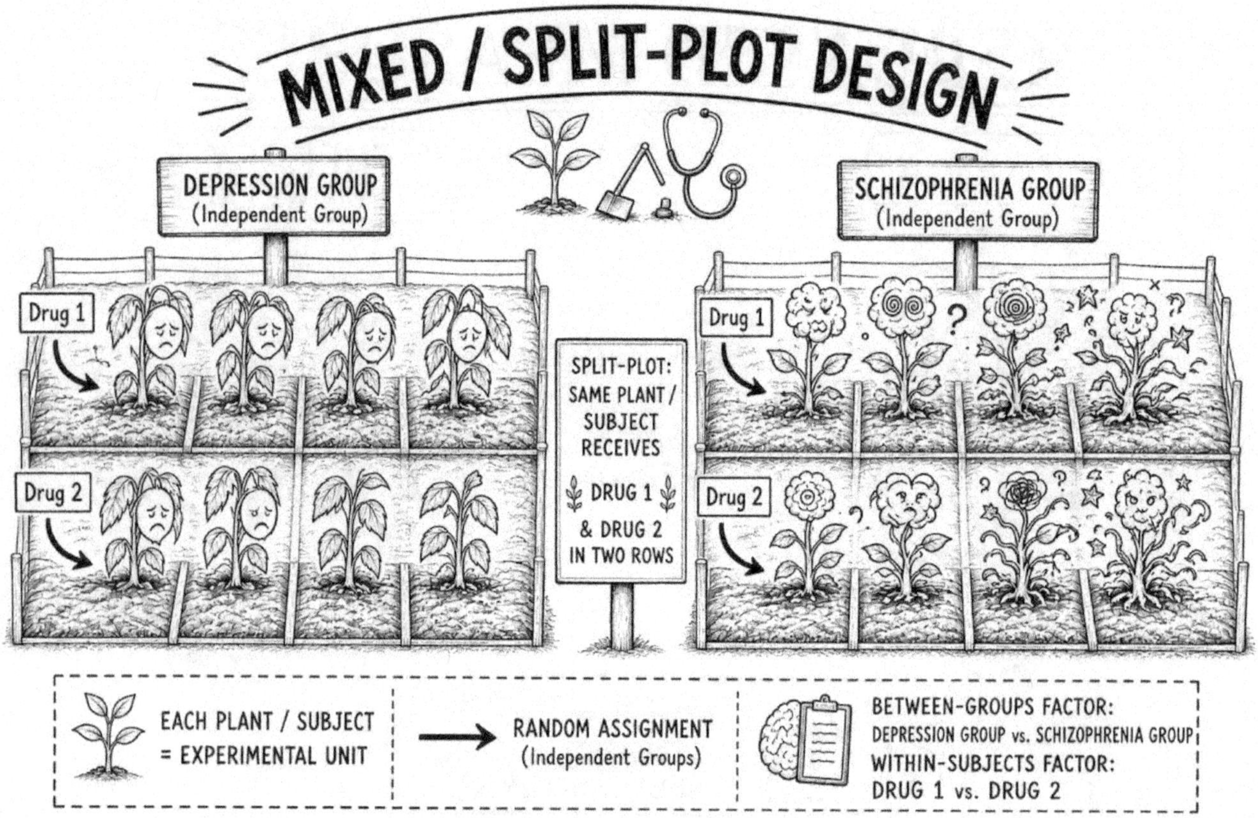

Figure 33. Mixed split-plot design

A mixed, or split-plot, design includes an independent-groups factor and a repeated-measures factor. For example, diagnosis may define independent groups, while drug condition is repeated within each subject. Different effects are tested against different error terms.

Four statistical ideas in words

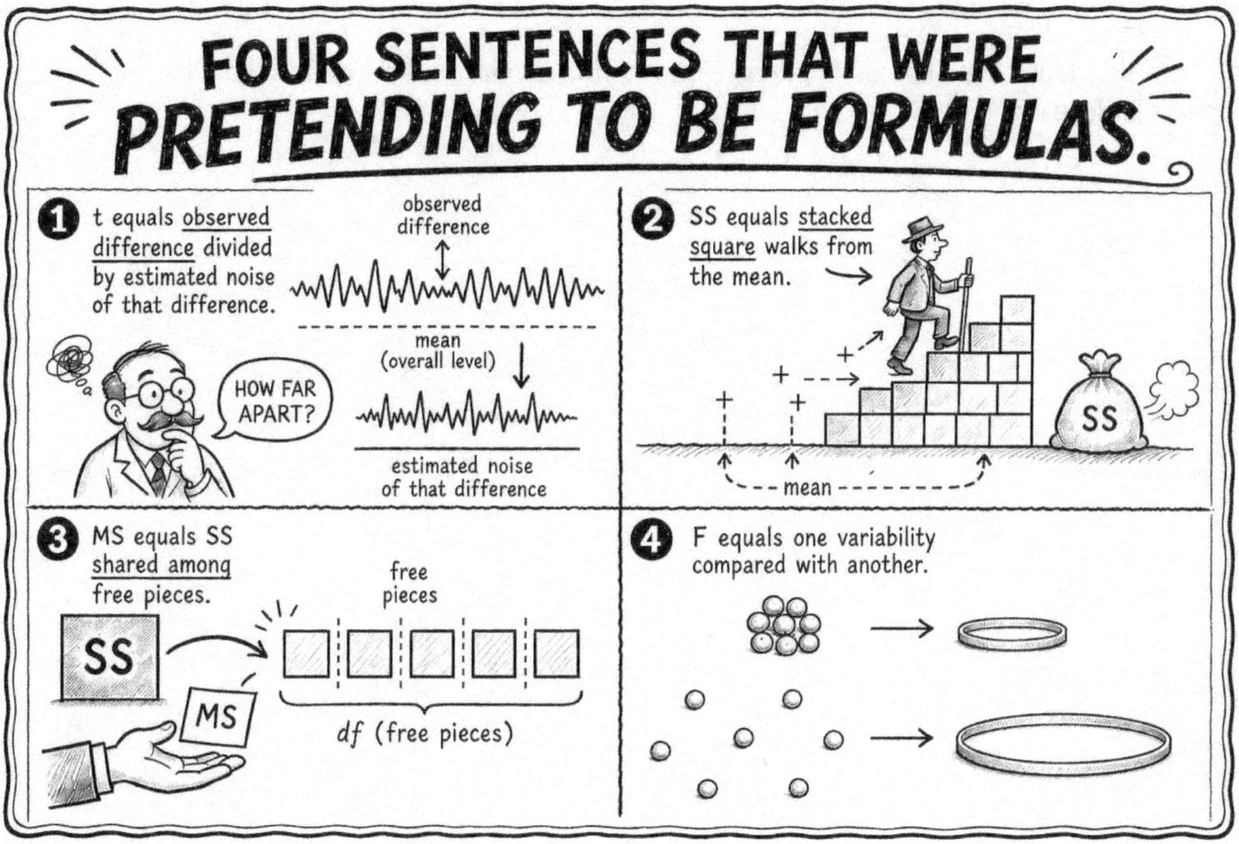

Figure 34. Four statistical ideas in words

$$t = \frac{\text{difference between sample means}}{\text{standard error of the difference}}$$

$$t = \frac{\bar{x}_1 - \bar{x}_2}{\text{SE}_{\bar{x}_1 - \bar{x}_2}}$$

$$t = \frac{\bar{x}_1 - \bar{x}_2}{s_p\sqrt{\frac{1}{n_1} + \frac{1}{n_2}}}$$

$$\text{sum of squares} = \text{sum of squared distances from the mean}$$

$$\text{SS} = \sum_n (x - \bar{x})^2$$

$$\text{SS} = \sum_{i=1} (x_i - \bar{x})^2$$

$$\text{mean square} = \frac{\text{sum of squares}}{\text{degrees of freedom}}$$

$$\text{MS} = \frac{\text{SS}}{df}$$

$$\text{MS} = \frac{\text{SS}}{n-1}$$

$$F = \frac{\text{variability between groups}}{\text{variability within groups}}$$

$$F = \frac{MS_B}{MS_W}$$

$$F = \frac{\frac{SS_B}{k-1}}{\frac{SS_W}{N-k}}$$

These are the main quantities used in basic parametric statistics. Understand what each quantity compares before using software to calculate it.

Part V Correlation

A third variable

Figure 35. Correlation is not causation

A scatterplot displays two measurements for each subject. An upward pattern shows a positive correlation, and a downward pattern shows a negative correlation. A patternless cloud indicates little linear correlation. The correlation coefficient, r, describes the direction and strength of the linear relationship. A correlation does not prove that one variable causes the other. For example, ice-cream sales and sunburns both increase in summer.

$$\text{correlation} = \frac{\text{sum of paired cross products}}{\sqrt{\text{product of the two sums of squares}}}$$

$$r = \frac{\sum(x-\bar{x})(y-\bar{y})}{\sqrt{\sum(x-\bar{x})^2\sum(y-\bar{y})^2}}$$

$$r = \frac{\sum_{i=1}^{n}(x_i-\bar{x})(y_i-\bar{y})}{\sqrt{\left[\sum_{i=1}^{n}(x_i-\bar{x})^2\right]\left[\sum_{i=1}^{n}(y_i-\bar{y})^2\right]}}$$

Summary

Figure 36. Summary of the statistical process

Identify the level of measurement and organize the data. Calculate an appropriate measure of center and variability, and draw a graph. Distinguish an observed distribution from the ideal normal curve. Use the t distribution when population variability is estimated from a sample. Draw the experimental design before selecting a test. Interpret t, F, SS, and MS as statistical comparisons, and do not treat correlation as proof of causation.

A teacher can show a cartoon without its caption and ask students to identify the statistical idea. If they can explain the action, they understand the concept before seeing the formula.

Epilogue

I hope this book has made the main statistical ideas easier to understand. These concepts can be useful in coursework, research, and professional practice.

I also hope that it encourages careful thinking about evidence, measurement, and uncertainty.

A formula is easier to use when the reader first understands the idea it represents.

About the author

Michael M. Nikoletseas, Ph.D., is a retired professor of Medicine and Neuroscience. He taught laboratory courses and statistics for decades, and he has published experimental work in psychology, neuroanatomy, and neurophysiology. A list of his books and research articles may be found in the catalogues of the Harvard and Oxford University libraries, and his teaching lineage may be found at neurotree.org.

This book accompanies his statistics textbooks. Those books explain the same concepts with text, examples, and tables; this volume introduces them through cartoons.

Statistics books by Michael M. Nikoletseas

Statistics for College Students and Researchers: Grasping the Concepts. 2010. ISBN 978-1453604533.

Statistics: Concepts and Examples. 2014. ISBN 978-1500815684.

Statistics for College Students and Researchers. Second Edition. 2020. ISBN 979-8588983235.

Statistics for College Students and Researchers. Third Edition. 2024. ISBN 979-8882788888.

Statistics for High School Students: Pre-College. 2024; revised August 2026. ISBN 979-8883980571.

Statistics and Data Science: Modern College Introduction. 2025. ISBN 979-8262744763.

Statistics Without Formulas: A Cartoon Book. 2026. The present volume.